by Anastasia Suen

ANIMAL BABIES

BEST EVER ANIMALS

Look for these words and pictures as you read.

kit

cub

chick

calf

Animal babies are the best!
They come in all sizes.
Some babies are big.

kit

Some babies are small.
A baby rabbit is a kit.
It is shorter than a crayon.

A baby bird is a chick.
This one is three crayons tall.

chick

cub

A baby bear is a cub.
It is as big as a dog.

calf

A baby giraffe is a calf.
It is taller than a teacher.

Animal babies come in all sizes. They are the best!

A baby cat is a kitten.
It is as light as an apple.

chick

calf

Did you find?

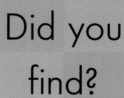

kit

cub

SPOT

Spot is published by Amicus Learning, an imprint of Amicus
P.O. Box 227, Mankato, MN 56002
www.amicuspublishing.us

Copyright © 2025 Amicus.
International copyright reserved in all countries.
No part of this book may be reproduced in any form
without written permission from the publisher.

Library of Congress Cataloging-in-Publication Data
Names: Suen, Anastasia, author.
Title: Animal babies / by Anastasia Suen.
Description: [Mankato, MN] : Amicus Learning, [2024] |
Series: Best ever animals | Audience: Ages 4-7 | Audience: Grades K-1 | Summary: "A search-and-find book about baby animals reinforces new vocabulary to build reading success while close-up images of baby animals captivate young audiences. A great early STEM book to inspire learning about animals and life science for kindergartners and first graders"—Provided by publisher.
Identifiers: LCCN 2023038575 (print) | LCCN 2023038576 (ebook) | ISBN 9781645492528 (library binding) | ISBN 9781681527765 (paperback) | ISBN 9781645493402 (pdf)
Subjects: LCSH: Animals—Infancy—Juvenile literature.
Classification: LCC QL763 .S84 2024 (print) | LCC QL763 (ebook) | DDC 591.3/92—dc23/eng/20231204
LC record available at https://lccn.loc.gov/2023038575
LC ebook record available at https://lccn.loc.gov/2023038576

Printed in China

Rebecca Glaser, editor
Deb Miner, series designer
Emily Dietz, book designer
  and photo researcher

Photos by Adobe Stock/geoffkuchera, 2, 8-9, 15; Alamy/Natalia Pryanishnikova, 14, Stefan Huwiler / Rolfnp, 2, 10-11, 15, Will Burrard-Lucas, 3; Deposit Photos/nawin_nachiangmai, 2, 4-5, 15; iStock/EEI_Tony, 12-13, GlobalP, 2, 7, 15, ouh_desire, Cover, 16, stefan1234, 1

ANIMAL BABIES